Pedestrian Safety in Australia

PUBLICATION NO. FHWA-RD-99-093

DECEMBER 1999

U.S. Department of Transportation
Federal Highway Administration

Research, Development, and Technology
Turner-Fairbank Highway Research Center
6300 Georgetown Pike
McLean, VA 22101-2296

FOREWORD

Creating improved safety and access for pedestrians requires providing safe places for people to walk, as well as implementing traffic control and design measures which allow for safer street crossings. A study entitled "Evaluation of Pedestrian Facilities" involved evaluating various types of pedestrian facilities and traffic control devices, including pedestrian crossing signs, marked versus unmarked crosswalks, countdown pedestrian signals, illuminated pushbuttons, automatic pedestrian detectors, and traffic calming devices such as curb extensions and raised crosswalks. The study provided recommendations for adding sidewalks to new and existing streets and for using marked crosswalks for uncontrolled locations. The "Evaluation of Pedestrian Facilities" also included synthesis reports of both domestic and international pedestrian safety research. There are five international pedestrian safety synthesis reports; this document compiles the most relevant research from Australia.

This synthesis report should be of interest to State and local pedestrian and bicycle coordinators, transportation engineers, planners, and researchers involved in the safety and design of pedestrian facilities within the highway environment.

Michael F. Trentacoste
Director, Office of Safety
Research and Development

NOTICE

This document is disseminated under the sponsorship of the Department of Transportation in the interest of information exchange. The U.S. Government assumes no liability for its contents or use thereof. This report does not constitute a standard, specification, or regulation.

The U.S. Government does not endorse products or manufacturers. Trade and manufacturer's names appear in this report only because they are considered essential to the object of the document.

Technical Report Documentation Page

1. Report No. FHWA-RD-99-093	2. Government Accession No.	3. Recipient's Catalog No.
4. Title and Subtitle **PEDESTRIAN SAFETY IN AUSTRALIA**		5. Report Date
		6. Performing Organization Code
7. Author(s) Peter Cairney		8. Performing Organization Report No.
9. Performing Organization Name and Address **AARB Transport Research Melbourne** — University of North Carolina Highway Safety Research Center, 730 Airport Rd, CB #3430, Chapel Hill, NC 27599-3430		10. Work Unit No. (TRAIS)
		11. Contract or Grant No. DTFH61-92-C-00138
12. Sponsoring Agency Name and Address **Federal Highway Administration Turner-Fairbanks Highway Research Center 6300 Georgetown Pike McLean, VA 22101-2296**		13. Type of Report and Period Covered
		14. Sponsoring Agency Code

15. Supplementary Notes

Prime Contractor: University of North Carolina Highway Safety Research Center
FHWA COTR: Carol Tan Esse

16. Abstract

This report was one in a series of pedestrian safety synthesis reports prepared for the Federal Highway Administration (FHWA) to document pedestrian safety in other countries. Reports are also available for:

United Kingdom (FHWA-RD-99-089)
Canada (FHWA-RD-99-090)
Sweden (FHWA-RD-99-091)
Netherlands (FHWA-RD-99-092)

Australia is a federation of States and Territories, and government responsibilities broadly mirror that in the USA. Local government is responsible for 80 percent of the road network, though the less heavily traveled parts. Australia is highly urbanized (notwithstanding large tracts of sparsely populated land). Almost 40% of the population lives in Melbourne or Sydney, and another 20% in Brisbane, Perth, and Adelaide.

Australia has been a pioneer of traffic calming in the form of Local Area Traffic Management, particularly in residential neighborhoods. Innovations are evident in the traffic signal area. Puffin crossings with infrared detectors seem promising. Pelican crossings are likely to find ready application, and having them set up for double cycle operations appears to offer benefits.

Australia was particularly innovative in developing the "safe routes to school" program, which integrates education, route selection, and engineering treatments to increase pupil safety. Also in development is the "walk with care" program designed for the elderly.

17. Key Words: **Australia, pedestrian crossings, local area traffic management, pedestrian safety, pedestrian signals**	18. Distribution Statement		
19. Security Classif. (of this report) **Unclassified**	20. Security Classif. (of this page) **Unclassified**	21. No. of Pages **40**	22. Price

Form DOT F 1700.7 (8-72)   Reproduction of form and completed page is authorized

SI* (MODERN METRIC) CONVERSION FACTORS

APPROXIMATE CONVERSIONS TO SI UNITS

Symbol	When You Know	Multiply by	To Find	Symbol
LENGTH				
in	inches	25.4	millimeters	mm
ft	feet	0.305	meters	m
yd	yards	0.914	meters	m
mi	miles	1.61	kilometers	km
AREA				
in^2	square inches	645.2	square millimeters	mm^2
ft^2	square feet	0.093	square meters	m^2
yd^2	square yards	0.836	square meters	m^2
ac	acres	0.405	hectares	ha
mi^2	square miles	2.59	square kilometers	km^2
VOLUME				
fl oz	fluid ounces	29.57	milliliters	mL
gal	gallons	3.785	liters	L
ft^3	cubic feet	0.028	cubic meters	m^3
yd^3	cubic yards	0.765	cubic meters	m^3

NOTE: Volumes greater than 1000 l shall be shown in m^3.

Symbol	When You Know	Multiply by	To Find	Symbol
MASS				
oz	ounces	28.35	grams	g
lb	pounds	0.454	kilograms	kg
T	short tons (2000 lb)	0.907	megagrams (or "metric ton")	Mg (or "t")
TEMPERATURE				
°F	Fahrenheit temperature	5(F-32)/9 or (F-32)/1.8	Celsius temperature	°C
ILLUMINATION				
fc	foot-candles	10.76	lux	lx
fl	foot-Lamberts	3.426	candela/m^2	cd/m^2
FORCE and PRESSURE or STRESS				
lbf	poundforce	4.45	newtons	N
lbf/in^2	poundforce per square inch	6.89	kilopascals	kPa

APPROXIMATE CONVERSIONS FROM SI UNITS

Symbol	When You Know	Multiply by	To Find	Symbol
LENGTH				
mm	millimeters	0.039	inches	in
m	meters	3.28	feet	ft
m	meters	1.09	yards	yd
km	kilometers	0.621	miles	mi
AREA				
mm^2	square millimeters	0.0016	square inches	in^2
m^2	square meters	10.764	square feet	ft^2
m^2	square meters	1.195	square yards	yd^2
ha	hectares	2.47	acres	ac
km^2	square kilometers	0.386	square miles	mi^2
VOLUME				
mL	milliliters	0.034	fluid ounces	fl oz
L	liters	0.264	gallons	gal
m^3	cubic meters	35.71	cubic feet	ft^3
m^3	cubic meters	1.307	cubic yards	yd^3
MASS				
g	grams	0.035	ounces	oz
kg	kilograms	2.202	pounds	lb
Mg (or "t")	megagrams (or "metric ton")	1.103	short tons (2000 lb)	T
TEMPERATURE				
°C	Celsius temperature	1.8C+32	Fahrenheit temperature	°F
ILLUMINATION				
lx	lux	0.0929	foot-candles	fc
cd/m^2	candela/m^2	0.2919	foot-Lamberts	fl
FORCE and PRESSURE or STRESS				
N	newtons	0.225	poundforce	lbf
kPa	kilopascals	0.145	poundforce per square inch	lbf/in^2

*SI is the symbol for the International System of Units. Appropriate rounding should be made to comply with Section 4 of ASTM E380.

(Revised September 1993)

TABLE OF CONTENTS

                                                                  Page

1. INTRODUCTION 1
    1.1   Australia 1
    1.2   Data Systems 1

2. SUMMARY OF PEDESTRIAN ACCIDENT EXPERIENCE 2
    2.1   Crash Trends 3
    2.2   Costs of road crashes 3
    2.3   The nature of the pedestrian problem in Australia 4
    2.4   Pedestrian exposure and over-representation 6
    2.5   The young and the elderly 7

3. OVERVIEW OF ACCIDENT COUNTERMEASURES AND SAFETY
       PROGRAMS 8
    3.1   Source Documents 9
            3.1.1   Australian Standard AS 1742.10- 1990
                     Manual of Uniform Traffic Control Devices, Part 10: Pedestrian
                     Control and Protection 9
            3.1.2   Austroads Guide to Traffic Engineering Practice, Part 13: Pedestrians 10
            3.1.3   Proposed Australian Road Rules 11
            3.1.4   Empirical studies of the safety of pedestrian facilities 13

4. CROSSWALKS 14

5. SIDEWALKS 14

6. SIGNALIZATION 14

7. SIGNING 15

8. MIDBLOCK CROSSINGS 15

9. MEDIANS AND PEDESTRIAN REFUGE AREAS 18

10. PROVISION FOR THE DISABLED PEDESTRIAN 18

11. SCHOOL ZONE SAFETY 19

12. PEDESTRIAN OVERPASSES AND UNDERPASSES 20

13. TRAFFIC CALMING FOR PEDESTRIANS 21
    13.1   Local Area Traffic Management 21
    13.2   Effect of humps and raised platforms 21

	13.3	Roundabouts (traffic circles)	22

14. INNOVATIVE DEVICES 23
    14.1 Infra-red sensors 23
    14.2 Pelican crossings 23
    14.3 Illuminated push buttons 24

15. OTHER ISSUES 24
    15.1 Multi-action programs 24
    15.2 Community Road Safety 25
    15.3 Pedestrian safety audits 26
    15.4 Speed and pedestrian safety 27
    15.5 Impact of lower speed limits on pedestrian safety 27
    15.6 The impact of seat belt wearing legislation on pedestrian safety in Australia 28

16. EDUCATIONAL CONSIDERATIONS 28
    16.1 Walk with care 30

17. ENFORCEMENT AND REGULATION 31

18. SUMMARY AND DISCUSSIONS 31

19. REFERENCES 32

1. INTRODUCTION

1.1 Australia

Australia is a federation of States and Territories in which the powers and responsibilities of State and Commonwealth Governments broadly mirror those of the State and the Federal Governments in the United States. As this applies to the roads sector, it means that each State has responsibility for managing the road system, determining the road laws that apply in that State, and recording information regarding traffic accidents. The Commonwealth Government has a role in funding the National Road Network and providing other funds to State and local governments to assist them in maintaining the road transport system.

The independence of the different jurisdictions leads to differences in traffic engineering practice and to differences in crash data bases which seriously limit the extent to which valid comparisons can be made among States. At present, a major effort is underway to develop a set of uniform road rules that all jurisdictions have agreed in principle to adopt. This should remove the few remaining minor differences among the States. The detailed implications for pedestrian safety are discussed below.

Local Government is responsible for approximately 80 percent of the road network in Australia. These tend to be the less heavily trafficked parts of the network. One important difference with the United States is that the local government's powers to enforce traffic regulations are generally confined to parking and weight restrictions with the traffic policing function being the responsibility of the State. Nevertheless, Local Government has a very useful role to play in improving road safety, particularly pedestrian safety.

Australia covers an area of just under 7.7 million km^2 (3 million mi^2), equivalent to 82 percent of the area of the United States. This land mass is occupied by just 18 million people, most of them located in a narrow band running along the eastern and southern coasts from Cairns to Adelaide. Australia is among the most urbanized countries in the world, with almost 40 percent of the population living in two cities — Sydney and Melbourne — and with a further 20 percent living in the major State capitals of Brisbane, Perth, and Adelaide.

Like that of the United States, Australia's population is extremely diverse. In 1992, 23 percent of the population had been born overseas, while data from the 1991 Census showed that 1.6 percent of the population was of Aboriginal or Torres Straight Islander descent (Australian Bureau of Statistics 1995). A feature of recent migration has been an increasing proportion of migrants from neighbouring countries in Asia.

1.2 Data Systems

Common with most other countries, crashes are reported by or to the police through the medium of a standard report form. Although there is a large degree of commonality, the Australian States differ in the specifics of what information is collected and how it is categorized. States also differ in their requirements to report crashes; for example, except in a few defined circumstances, Victoria requires an accident report only when there has been a personal injury, New South Wales requires a report only when there has been an injury or a vehicle is towed away from the scene of an accident, and most other States require a report when there has been an injury or the estimated damage exceeds a set amount. In practice, the different reporting requirements have little impact on the reporting of pedestrian crashes as nearly all of these involve a degree of personal injury. However, under-reporting of crashes is acknowledged as a problem. A recent Western Australian study of the match between police and hospital records had a match rate for pedestrians of 68.6 percent (Rosman and Knuiman 1994). This study confirmed many of the usual findings regarding non-reporting of crashes, such as non-reporting being more likely for accidents involving children and less serious injuries. New findings from this study are that match rate varied according to the type of hospital with the metropolitan teaching hospitals having the highest rates and private hospitals the lowest, and that matching rates vary according to ethnicity, with Aboriginal Australians and Asians having lower matching rates than non-Aboriginal Australians, and European-born persons having higher matching rates.

Each State publishes an annual statement of road crashes that have occurred each year. These statements differ considerably in their level of detail and sophistication. The New South Wales statistical statement is probably the most comprehensive and has been issued in its current form for more than 10 years (eg RTA 1994). However, even this publication gives relatively little information specifically about pedestrian crashes — number of crashes broken down by Road User Movement (RUM) code and numbers of pedestrian casualties broken down by severity of injury, sex, and age.

One notable feature of the reporting systems in a number of States is the very detailed coding of crash events, referred to as RUM Codes, or Definitions for Classifying Accidents (DCAs). Crashes are allocated to one of many types of event, defined in terms of a diagram that encapsulates the essential features of the actions of the road users involved in the crash. Similar systems are currently used in New South Wales, Victoria, and Queensland and are under consideration by some of the other States. Australian experience has shown that having this type of classification available in the data base provides a powerful tool that greatly adds to the understanding of crash patterns, some examples of which will be discussed later in this paper.

2. SUMMARY OF PEDESTRIAN ACCIDENT EXPERIENCE

Pedestrian deaths and hospital admissions for Australia in 1993 are shown in table 1. Pedestrians constitute 17 percent of road fatalities nationally, the proportion ranging from 12 percent to

Table 1. Fatalities, hospital admissions and crashes involving pedestrians and all road users, 1993.

	Pedestrians	All Road Users
Fatal crashes	327 (19%)	1734
Fatalities	331 (17%)	1953
Hospital admission crashes	2557 (15%)	17186
Hospital admissions	2681 (12%)	21602

20 percent across the States and Territories. Pedestrians account for 12 percent of hospital admissions resulting from road crashes, ranging from 10 percent to 15 percent across States, with slightly lower proportions in the two territories.

These crashes represent a pedestrian fatality rate of 1.84 per 100,000 persons. This compares with a U.S. pedestrian fatality rate of 2.13 per 100,000 persons for the country as a whole (NHTSA 1996), and a crash rate for Great Britain of 1.83 per 100,000 persons (Department of Transport 1996). Australia's pedestrian fatality rate is equivalent to the middle-ranking States when U.S. states are ranked by pedestrian crash rates (see NHTSA 1996, chapter 5).

2.1 Crash Trends

Since national serious injury data is available for different road user classes only from 1989 onwards, the discussion of crash trends is restricted to this period. As figure 1 shows, pedestrian fatalities have declined from a high of 501 in 1989 to 330 in 1993, a fall of 34 percent. All road fatalities have fallen by 30 percent, from 2,803 to 1,953, over the same period.

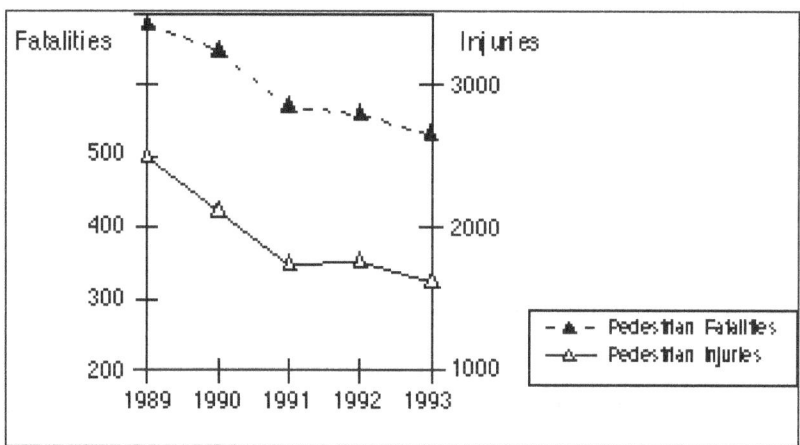

Figure 1. Trends in pedestrian fatalities and injuries resulting in hospital admission, Australia 1989-1993.

Pedestrian hospital admissions have declined by 23 percent from 3,478 to 2,681, and serious injuries to all classes of road users have declined by 24 percent from 28,490 to 21,602. Thus reductions in deaths and injuries to pedestrians are consistent with reductions in deaths and injuries for all road users.

2.2 Costs of road crashes

Andreassen (1992) made comprehensive estimates of the cost of road crashes, taking into account the costs of emergency service attendance, medical and hospital costs, loss of earnings, grief, pain and suffering, legal and administrative costs, and vehicle repair costs. One significant feature of this work is its recognition that different types of crash have different outcomes in terms of injuries and vehicle repair costs, and hence costs were calculated for each accident-type separately for urban and rural crashes.

Estimates of the cost of pedestrian crashes were based on data from three types of pedestrian crashes only — the near side, far side, and emerging crash types — and were based on data from New South Wales and Victoria. However, these three types account for the bulk of pedestrian crashes. The estimated cost of a pedestrian crash in metropolitan areas was $A79,300 ($51,545), and in rural areas $A148,800 ($96,720). The latter figure reflects higher speeds in rural areas, with the consequent greater probability of serious death or injury. Only head-on crashes were estimated to cost more than pedestrian crashes. With the exception of head-on crashes, the estimated range of other types of crashes was approximately $A18,000 ($11,700) to $A70,000 ($45,500) for metropolitan crashes and $A23,000 ($14,950) to $A107,000 ($69,550) for rural crashes.

2.3 The nature of the pedestrian problem in Australia

Cairney and Cusack (1997) compared pedestrian crashes in New South Wales, Victoria, and Queensland. These three States use similar procedures for coding crashes, based on the approach used in New South Wales and Victoria for several years now, and further clarified and refined by Andreassen (eg Andreassen 1994). In this approach, a crash is classified as belonging to a particular crash type according to the pattern of road user movements preceding the first collision between vehicles or between a vehicle and a pedestrian. In all, there are some 88 categories to which crashes can be allocated. Of these, nine relate specifically to pedestrian crashes.

Cairney and Cusack found a high degree of consistency among the States in all factors relating to pedestrian crashes, including crash types. These results are summarised in table 2. In all three States, the most common type of crash involved pedestrians being hit on the near side of

Table 2. Pedestrian accidents by accident type (from Cairney and Cusack 1997).

Accident Type Code & Description		Victoria		NSW		Queensland	
		No.	%	No.	%	No.	%
001	Pedestrian — Near side vehicle hit from right	1,477	41.6%	2,887	47.1%	686	46.8%
002	Pedestrian — Hit emerging behind vehicle	394	11.1%	874	14.2%	101	6.9%
003	Pedestrian — Far side vehicle hit from left	1,006	28.4%	1,531	25.0%	442	30.2%
004	Pedestrian — Playing, working, lying, standing, on carriageway	253	7.1%	330	5.4%	105	7.2%
005	Pedestrian — Walking with traffic	107	3.0%	170	2.8%	55	3.8%
006	Pedestrian — Facing traffic	43	1.2%	48	0.8%	36	2.5%
007	Pedestrian — Hit on footpath/ median	68	1.9%	170	2.8%	16	1.1%
008	Pedestrian — Hit on driveway	108	3.0%	126	2.1%	19	1.3%
009	Pedestrian — Struck while boarding or alighting vehicle	91	2.6%	0	0.0%	6	0.4%
	Total	3,547	100%	6,136	100%	1,466	100%

the carriageway by a vehicle approaching from the right. (Remember that traffic keeps to the left in Australia.) This type of crash accounted for almost half the pedestrian crashes. The second most common type of crash in all three States was where the pedestrian was hit on the far side of the carriageway by a vehicle approaching from the left. This type accounted for between 25 percent and 30 percent of pedestrian crashes. In New South Wales and Victoria, the next most common type of crash involved pedestrians emerging from behind a vehicle. This type of crash was less prevalent in Queensland, where it was nevertheless the fourth most frequent type of crash. The fourth most frequent type of crash in Victoria and New South Wales, and the third most frequent in Queensland, involved the pedestrian playing, working, lying, or standing on the carriageway. In all three States, these four categories accounted for over 80 percent of crashes. None of the other categories covered large numbers of crashes, but it is worth noting that slightly more crashes in Queensland involved walking with or facing the traffic, while crashes involving boarding or alighting from a vehicle were a much greater feature of the Victorian data, presumably as a result of patronage of Melbourne's extensive tram (streetcar) system.

In general, the proportions of different types of crashes are similar in the metropolitan and other urban areas. However, in rural areas there are fewer nearside, emerging, and far side crashes, but more crashes where the pedestrian has been walking with or against the traffic, and more in the category playing, walking, standing, or lying on the road.

A more detailed picture of the risk factors affecting pedestrians of different ages was produced by a very comprehensive study carried out by the Victorian Road Traffic Authority (RTA) in the early 1980's (Alexander, Cave, and Lyttle 1990). Persons admitted to hospital or treated as an outpatient as a result of being injured as a pedestrian in a collision with a motor vehicle were interviewed. They answered questions relating to trip purpose, circumstances of their crash and contributing factors, and their demographic characteristics. Their answers were then compared with the answers given by matched control subjects, who were asked similar questions, excluding of course the questions relating to accident involvement. The control subjects were interviewed at the same place that the pedestrian had been injured, and at the same time of day. Some of the interviewees had been matched to the injured pedestrian on the basis of age and sex, some not. This provides a powerful way of testing which factors affect risk among pedestrians of the same age group, and testing how much greater is the risk faced by a particular age group.

It was found that people aged 60 to 80 had double the crash risk of younger pedestrians. They were more likely to have been migrants, but no less likely than controls to have English as a first language, suggesting an over-representation of English-speaking migrants. They were no different from controls in terms of marital status, physical disability, or psychological state at the time of the crash. Most of the crashes happened in shopping areas, and 70 percent of victims were shopping or on the way home from shopping. A high proportion had been carrying shopping bags. Pedestrian behavior in these crashes was characterized by a high degree of compliance with traffic control devices and regulations, but an inability to complete crossings in time or to anticipate unexpected actions on the part of drivers.

One primary focus of the study was the role of alcohol in pedestrian crashes (see next section). No matched controls were sought for child pedestrian crash victims, and therefore it is not possible to be confident about which factors are associated with increased risk. However, it was worth noting that 67 percent of this group were males, and 91 percent were born in Australia. While these constitute over-representation in terms of the numbers of these individuals in the population, without the control data it is not possible to say whether this constitutes over-representation in terms of the pedestrians at that site and time. Most of the child pedestrian crashes occurred on the urban fringe, and 51 percent occurred during the period 4-6 p.m. Most of the crashes were mid-block crashes. Of those occurring at intersections, 57 percent had a local street forming at least one leg of the intersection. A child had been crossing the road in 88 percent of cases, 31 percent emerging from behind vehicles, and 20 percent may have spontaneously darted onto the road. Fifty-seven percent had their view obstructed, particularly by parked cars. The crashes were characterized by high traffic density, poor sight distance, distractions, and a lack of supervision.

This study also confirmed the major contribution of alcohol to pedestrian crash causation in Victoria. Forty percent of adult crash victims had been drinking, and 24 percent had blood alcohol concentrations (BACs) greater than 0.15. The increase in risk with increased BAC is very large for pedestrians, similar to the increases in risk experienced by motor vehicle drivers with similar increases in BAC. Pedestrians with a BAC greater than 0.10 were found to have double the risk of being involved in a crash compared with those who have a BAC of less than 0.10, while those who have a BAC greater than 0.15 have 15 times the risk of being involved in an injury crash of those with a BAC less than 0.10.

Victims who reported high levels of alcohol consumption appeared to be at greater risk. High BAC crashes (i.e., those where the pedestrian who had a BAC of 0.15 or greater) were associated with:

- Near-side crashes (67%).
- Weekends (52%).
- Occurring between 6 p.m. and 6 a.m. (78%).
- Within 400 m (1,312 ft) of a pub or other drinking venue (70%).
- Inner suburban areas (56%).

A more recent examination of Queensland's pedestrian crash data confirms that this general picture applies in that State (Fraine 1995).

2.4 Pedestrian exposure and over-representation

Anderson, Montesin, and Adena (1989) estimated the risk of travel by different modes, based on a survey of day-to-day travel patterns and carried out on behalf of the Federal Office of Road Safety. Data on fatal crashes were obtained from that organization's Fatal File for 1984 and 1985. The average number of fatalities per 10 million km (6 million mi) travel for males and females for different modes are shown in table 3 on the following page.

These data suggest that the risk of being killed while walking is about 12 to 15 times the risk of traveling by car per unit distance, and that walking has almost the same risk of being killed as does motorcycling. However, since walking is so much slower than other modes and the trip length typically much shorter, this is not a meaningful comparison. A different picture emerges

Table 3. Fatalities per 10 million km (6 million mi) travel by different modes, Australia 1984-5. Adapted from Anderson et al (1989)

Mode	Males			Females		
	per 10^8 km	per trip	per 10^6 hours	per 10^8 km	per trip	per 10^6 hours
Car drivers	0.15	0.19	0.53	0.10	0.08	0.31
Car passengers	0.17	0.27	0.66	0.09	0.13	0.34
Motorcyclists	2.86	3.01	9.77	1.48	1.01	23.65
Bicyclists	0.57	0.14	11.36	0.25	0.05	1.30
Pedestrians	2.35	0.26	1.09	1.26	0.12	0.46

when the risk of being killed per trip by the various modes is examined. The risk of being killed while making a journey by any of the modes is roughly the same, with the exception of motorcycling, which again emerges as having a much higher risk of being killed. A slightly different picture emerges when fatalities per million hours traveling by each mode are considered. Cycling emerges as having a very high fatality rate, especially for males, but male pedestrians also seem to have a high fatality rate.

In Australia, the different States and Territories have different legal reporting requirements and different numbers of hospitalized and less seriously injured victims in relation to the numbers killed. The reasons for this are not fully understood but may include reasons such as different availability of hospital beds, different insurance arrangements, and different reporting cultures (see e.g., Cairney and Cusack 1997). It is therefore not possible to estimate meaningful injury rates in the same way as it is possible to obtain fatality rates.

2.5 The young and the elderly

Cairney and Cusack (1997) examined the fatality rates for pedestrians and cyclists of different ages across three States. Although there was variation among the States, the rates tended to be high among 5 to 22 year olds, although in Queensland the rate was quite low among the 9 to 17 year olds, then very high among the 18 to 22 year olds, and remaining high among the 23 to 32 year olds. All three States had a high rate among the 58 to 67 year olds and an extremely high rate among those aged 68 and older. Cairney and Cusack also examined rates of hospital admissions and injuries requiring medical treatment. For the reasons discussed above, the injury rates are not directly comparable among States. However, when the data for each State are examined for each State independently, a common pattern emerges for both classes of injury which is similar to the pattern for fatalities. Rates are high between the ages of 5 and 22, then again in the group aged 68 and over.

Australian data confirms this picture of a large difference between the sexes and between age groups for both pedestrians and cyclists. Anderson, Mortesin, and Adena (1989) considered age and fatality rate per 10^7 km (6 mi) in their investigation of fatality rates. Examination of the raw tables indicates erratic trends. Anderson et al undertook statistical modeling of their data to produce regular curves that are readily interpretable. The most easily interpretable are those presented in the appendix of their report, in which the actual crash rates are modeled, rather than the logarithm of the fatality rate which is used throughout the report. The curves relating fatality rate for pedestrians to age are shown in figure 2.

Figure 2 shows four separate curves representing crash rates for males and females at different times of day. The main features of this figure are:

- Early night, late night, and evening rates are higher than day time rates.
- The rates for men are consistently higher than those for women.
- Daytime rates show little variation with age compared to other times.
- Children and teenagers have high fatality rates in the early night and late night periods.
- Older people, especially men, have extremely high fatality rates in the evening and early night.

It should be borne in mind that this data set does not include data from children under 9 years old.

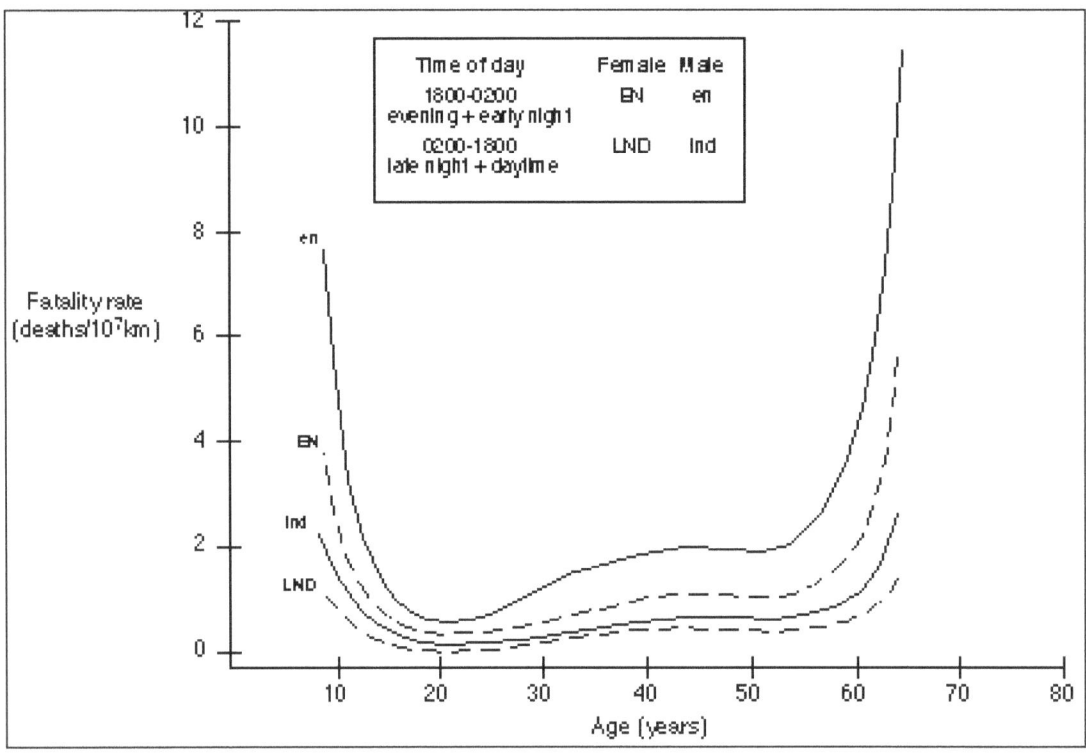

Figure 2. Total crash rates for male and female pedestrians of different ages at different times of day.

3. OVERVIEW OF ACCIDENT COUNTERMEASURES AND SAFETY PROGRAMS

With responsibility for operational traffic safety matters divided among eight geographically remote jurisdictions, it is difficult to give a comprehensive overview of accident countermeasures and safety programs for all of Australia.

However, it is probably a fair generalization that pedestrian safety initiatives have been focused primarily on engineering treatments and, to a lesser extent, education. In contrast to other areas of road safety improvement in recent years, such as reducing alcohol-impaired driving and speeding, enforcement has had very little impact. The reasons for this are discussed in section 17. While other classes of road users have benefited in recent years from improved personal protection, backed by legislation requiring the use of seat belts, baby capsules and child seats, and bicycle helmets, there has been no equivalent improvement in the protection available to pedestrians, apart from changes to vehicle front ends. Australian pedestrians may have benefited less from changes to vehicle design than pedestrians in other developed countries. The Australian vehicle fleet is relatively old by world standards, and a low base-line replacement rate for vehicles has been compounded by a sluggish economy in recent years.

3.1 Source Documents

There are three basic source documents that cover signs and markings for pedestrian facilities, provision and design of pedestrian facilities, and proposed legislative changes.

The Australian Standard *Manual of Uniform Traffic Control Devices* and the Austroads *Guide to Traffic Engineering Practice* provide a set of guidelines for the control and protection of pedestrians. In practice they are used as guidelines and are not legally enforceable. The draft Australian Road Rules has a number of sections relating to pedestrians. The Australian Road Rules will not be enforceable until legislated by Federal Parliament. It is expected that the Rules will then become an Australia-wide standard, replacing current traffic regulations in each of the states.

3.1.1 Australian Standard (AS) 1742.10 — 1990
Manual of Uniform Traffic Control Devices, Part 10: Pedestrian Control and Protection

- This standard sets out requirements for traffic control devices to be used in the control and protection of pedestrian traffic on roads. It specifies the way in which these are used to achieve pedestrian control. The manual includes definitions, installation details, clause references, and references to other applicable standards. Requirements for the illumination and reflectorization of signs, their installation, location, and size are outlined in the appendixes. Details are also included on model instructions for adult supervisors and child monitors at children's crossings, pedestrian-actuated traffic signals, and pedestrian treatments at railway level crossings.

- **People with disabilities**: People with a vision impairment have difficulty with visual cues and need a strong contrast/delineation between the road and pedestrian areas, usually a physical guide. Obstructions to their path, such as street furniture and sign posts, can cause difficulties. People with a hearing impairment will rely on seeing vehicles to cross safely and therefore need a clear view.

Wheelchair users need continuous, even, and hard surfaces. Ramps need to have a low gradient, and curb edges need to allow roads to be easily crossed.

- **Young children**: The size of young children limits their ability to see and be seen, and this needs to be considered when designing barriers and signs near crossings and on islands. For design purposes, children should not be considered as mini-adults as they do not have the cognition or perception to make sound judgments until approximately 12 years of age.

- **The elderly**: The elderly have lower walking speeds and reduced abilities, which are often not adequately catered for. Older people tend to be aware of their reduced abilities and so tend to adjust their behavior.

The classification and type of pedestrian facilities are as follows:
1. Time separation;
2. Physical (or spatial) separation; and
3. Integrated facilities (adequate control where integration of pedestrians with vehicles is acceptable).

Time separation
These are installed if the degree of hazard warrants imposing regulatory controls on the driver and occasionally on the pedestrian. Conflict is minimized by allowing short-time periods for exclusive pedestrian use of a specific section of road, alternated with vehicle use only of the same section. The treatments covered by this section include pedestrian crossings (zebras), children's crossings, pedestrian's actuated traffic signals (mid block), pelican crossings, and provisions for pedestrians at signalized intersections.

Physical (or spatial) separation/ Physical pedestrian aids
These devices are used to reduce the amount of exposure between pedestrians and vehicles. They may be warranted for wide, heavily trafficked roads, where there is not enough pedestrian traffic to justify a time separated device. This simplifies the decisions both groups have to make. Treatments covered here include pedestrian refuges, traffic islands and medians; curb extensions; loading islands, and safety zones; and pedestrian fencing.

Physically separated facilities can provide greater protection for pedestrians and minimum disruption to road traffic. Treatments include subways, bridges, and pedestrian malls.

Integrated facilities
These are applied where the presence of pedestrians or pedestrians with a special characteristic (e.g., school children) may share a road space in a largely unsupervised manner. Treatments include pedestrian warning signs, shared zones, school zones, local area traffic management schemes, and lighting.

3.1.2 Austroads Guide to Traffic Engineering Practice, Part 13: Pedestrians

This publication is a guide for designing pedestrian facilities, which takes into account the requirements of the various Australian standards and road design guidelines, and which gives detailed consideration to the requirements of groups with special needs. It assists in the interpretation and application of the different standards dependent on the situation. This ensures that the traffic control and safety schemes introduced are directly applicable to all potential users. Pedestrians are considered vulnerable road users and represent a significant portion of serious road injuries and fatalities. Their risks may be reduced with simple safety measures.

This document aims to remind designers that pedestrians have a different range of capabilities, behaviors, and attitudes. It particularly highlights important design factors to be considered for those who do not fit the normal pedestrian profile, such as the sight and mobility impaired, young children, and the elderly. It shows how these groups can be catered for in roadway environments encountered by pedestrian traffic.

When designing pedestrian facilities, the nature of pedestrian demand needs to be ascertained. This involves considering the flow and depth of pedestrians, journey origins and destinations, purpose of trip, and peak times. A strategy incorporating all aspects of pedestrian comfort, convenience, and safety can then be adequately developed. The Austroads document aims to assist designers with consolidation of practices used in different states and, where applicable, overseas. It is designed to be a source document for guidelines, standards, and practices, as well as recommendations. With this in mind Austroads provides advice and techniques to assist in identifying problem locations and appropriate engineering treatments for them in respect to pedestrian facilities. Guidelines are provided on the appropriate standards for walkways and footpaths, the provision of pedestrian facilities for crossing roads, signing and other guidance methods, and treatments applicable for public transport, work sites, and parking areas.

3.1.3 Proposed Australian Road Rules

Under Australia's constitutional arrangements, each State has responsibility for land transport within its borders. This includes legislation governing traffic matters. Considerable advances have been made in recent years in harmonizing regulations and standards across the nation. Agreement is close on the Australia Road Rules which, it is intended, will supersede existing State legislation by having each State pass a new act embodying the Australian Road Rules.

At present, there remain many differences among the regulations in different States, although major incompatibilities have been removed and drivers would appear to have few difficulties with changes in regulations when driving outside their home State.

The current draft document defines a pedestrian as a person on foot; a person traveling on or in a pram, a wheeled toy; a person walking a bicycle, motorbike, or animal; a person traveling in a non-motorized wheelchair or a wheelchair that can not exceed 10 km/h (6 mi/h) and is less than 110 kg (242 lbs). The document refers to pedestrians in relation to obeying traffic lights (part 4); giving way to pedestrians, vehicles, and animals (part 6); and special rules applying to pedestrians (part 13).

The following is specified in relation to pedestrian movement:

Obeying Traffic Lights

4.10(1) Motorists must give way to pedestrians on a marked foot crossing that is not at an intersection when the traffic light that controls it is flashing yellow.

4.12(1) Pedestrians should not cross a road if there is a red "don't walk" sign, there is a red walking man symbol, there is no pedestrian light showing, or there are twin flashing red traffic lights. Section 4.12(2) outlines some exceptions to this rule.

4.13(1) Pedestrians must cross to the nearest safety zone, dividing strip, or edge of road if any signal illustrating that they should not cross is displayed.

Giving Way to Pedestrians, Vehicles, and Animals

6.2(1)/6.3(1) A driver must approach a children's crossing or pedestrian crossing at a speed where the vehicle can be stopped safely before reaching it.

6.2(2)/6.3(2) If a pedestrian is on the children's crossing, or a pedestrian crossing, the motorist must stop before reaching it and give way.

6.4(1) A motorist must not enter a children's crossing, pedestrian crossing, or marked foot crossing if a vehicle facing in the same direction appears to be stopped at the crossing.

A motorist must give way to pedestrians in the following circumstances:

6.5(1) In a shared zone.

6.6(1) Entering or leaving a road where a pedestrian is traveling across the road, footpath, bicycle path, separated footpath, or nature strip that must be crossed to get to or from that road.

6.7(3) When a pedestrian is crossing a road that a motorist is turning right into where an intersection is not controlled by traffic lights or signals.

6.7(4) When a pedestrian is crossing a road that a motorist is leaving or intending to enter when turning left at an intersection that is not controlled by traffic lights or signs.

6.7(5) When a pedestrian is crossing a left-turn slip lane that a motorist is turning left on.

6.8(1) At a T-intersection that is not controlled by traffic lights, a motorist wishing to turn into the continuing road must give way to any pedestrian crossing it.

6.8(2)/6.8(3) At a T-intersection that is not controlled by traffic lights, a motorist about to leave the continuing road must give way to any pedestrian crossing the road that ends at the intersection.

6.9(3) If at a green or flashing yellow traffic light and turning left, the motorist must give way to any pedestrian crossing the road they are turning into.

Special Rules Applying to Pedestrians

13.2(1) A pedestrian must not move into the path of an oncoming vehicle if that would cause a traffic hazard.

13.3(1) A pedestrian must not cross the road if it is within 20 m (66 ft) of a marked foot crossing, a pedestrian crossing, or a children's crossing. Where this does not apply is covered in 13.3(2).

13.4(1) A pedestrian must cross a road in the shortest most direct way. Where this does not

	apply is covered in 13.4(2).
13.5(1)	A pedestrian must cross a road as quickly as possible.
13.5(2)	A pedestrian waiting for a tram or public bus must not cross the road until the tram or bus is stopped.
13.7(1)	A pedestrian must not walk on a road if able to walk on a path.
13.7(2)	If walking on a road, a pedestrian must walk as close to the edge as possible.
13.7(3)	Pedestrians must not walk more than two abreast on a road.
13.8(1)	In a shared zone, a pedestrian must not obstruct the path of another pedestrian or vehicle.
13.9(1)	A pedestrian crossing a railway line must use the means provided.
13.10(1)	Pedestrians may not get on a moving vehicle; remain on any road crossing longer than is necessary to cross, walk on a path marked for bicycles, or obstruct a cyclist when crossing a path designated for bicycles.
13.10(3)	A person in a wheelchair may travel on a path designated for bicycles.

3.1.4 Empirical studies of the safety of pedestrian facilities

Moore and McLean (1995) carried out a review of pedestrian facilities for the South Australian Department of Transport. The first part of that report raises general issues related to the provision and operation of pedestrian facilities (dealt with elsewhere in the present report), while the second part deals with specific types of facilities.

Geoplan (1994) arrived at estimates of the relative safety of different types of devices by a different route. In a study commissioned for the New South Wales Roads and Traffic Authority (Road Transport Authority), Geoplan examined traffic flow and accident data for many sites where a range of different types of traffic facilities had been installed over the period 1985-1993 and compared crash rates before and after the installation of the pedestrian facilities. Since many of these facilities were installed on non-arterial roads belonging to local authorities, these local authorities had to be contacted by questionnaire to identify the range of sites. Sites where traffic signals had been installed could be identified from RTA records. The study was confined to the metropolitan centers of Sydney, Wollongong, Newcastle, and their vicinity.

There is one great advantage in a before-after comparison of this type. The treated and untreated sites are the same sites at different times. One of the difficulties in comparing treatment sites with control sites is that it is always difficult to identify suitable control sites. Studies of driver behavior have made it clear that subtle differences in site characteristics can result in greater changes in behavior than do even apparently powerful treatments. Using the same sites as the treated and untreated sites, eliminates these differences.

However, there are several unavoidable problems with the study. Some of the comparisons are based on small numbers of sites and therefore have insufficient statistical power. The before and after periods differ from device to device, forcing a reliance on crash rates over time for analysis. Many of the after periods are short, further weakening the power of the comparison. There may have been changes in pedestrian exposure over the period which are not taken into account for the study.

One possibility is crash migration, i.e., changes in travel patterns or driver behavior as a result of installing the facility, and more crashes occurring elsewhere as a result.

4. CROSSWALKS

AS 1742.10 describes the type of crosswalk to be provided at mid-block and intersection signals, and only at these locations. Although AS 1742.10 does not require a marked crosswalk, it is general practice to provide one. The crosswalk lines may be broken or continuous, according to

State practice. The installation of crosswalks has not been a contentious issue in Australia, and there is no research on their effectiveness.

5. SIDEWALKS

Sidewalks, referred to as footpaths in Australia, are generally provided in urban areas. A general minimum width of 1.2 m (4 ft) is specified, with an absolute minimum of 0.9 m (3 ft) (Austroads 1995). Wider footpaths are called for if pedestrian volumes are large, or if provision is required to be made for wheel chairs, or if the facility is to be shared with cyclists. The Austroads guide contains guidelines for the design details and the characteristics of suitable surfacing materials.

6. SIGNALIZATION

In Australia, pedestrian signals have been provided at both mid-block locations and at intersections. AS 1742.10 gives guidelines for the installation of both types of facilities. For the mid-block pedestrian operated signals, these include special provision for schools, danger to pedestrians or a history of pedestrian accidents at the site, and excessive delays to road traffic, as well as combinations of vehicle and pedestrian movements.

The pedestrian displays used in Australia consist of a standing steady red figure to indicate "Do not cross," a steady green figure in walking pose to indicate "Cross," followed by the red figure in flashing mode to indicate the pedestrian clearance phase. Dissatisfaction is frequently expressed with the flashing red figure as many pedestrians, particularly the elderly, seem to interpret it as meaning "Hurry and complete your crossing" rather than "Do not commence crossing." Cairney (1988) used video sequences to test whether substituting a flashing green figure to indicate the clearance phase would convey the message more clearly, but found no difference between the flashing red figure and the flashing green. More recently, Catchpole et al (1996) used observations of pedestrian behavior and interviews with pedestrians to test whether a flashing yellow figure gave rise to a better understanding of the intended message during the clearance phase, but found that it conveyed no obvious advantage.

The essential requirements for signal timing for pedestrians are also given in AS 1742.10. The minimum pedestrian green time specified is 6 s, and the pedestrian clearance phase is calculated to give a pedestrian time to complete a crossing of the road, assuming a walking speed of 1.2 m/s (4 ft/s), with a minimum of 5 s.

The Geoplan study found that the provision of pedestrian signals at existing facilities did not reduce pedestrian crashes (Geoplan 1994). Although the crash rate per quarter declined by 22 percent following the installation of pedestrian signals, this was less than the crash rate fell at other similar sites where no facilities had been installed. When the correction was made, crashes were estimated to have increased by 1 percent as a result of the installation. This, of course, was not statistically significant.

7. SIGNING

AS 1742.10 (standards Australia, 1990) follows the American version road signs ratified by the 1968 United Nations Convention, and therefore follows the shape and colour code used in North America. AS 1742 differs from American practice in that it relies to a much greater extent on symbol signs; the diamond and rectangular shapes specified for signs allowing for a large symbol with good legibility distance. The principal signs used for pedestrian protection are shown in figure 3.

The pedestrian crossing sign is something of an anomaly, being a survivor from a much older version of the traffic sign standard. It is currently the only circular sign in AS 1742, and is yellow with black symbol and lettering, a color combination otherwise reserved for warning or roadwork signs. The sign warning of pedestrians is virtually identical to the one currently in use in North America and is also yellow with a black image. Not surprisingly, there is some confusion over the regulatory or warning nature of the signs (Cairney 1989), although this is unlikely to make much difference to driver behavior towards pedestrians, especially since the Pedestrian Crossing sign is always provided in conjunction with a striped zebra crossing. Note that there is also a symbolic warning sign for children and a symbol sign warning of a pedestrian crossing ahead.

Other signs of particular importance for pedestrian safety include the written Safety Zone, School and School Zone signs, and the hybrid Shared Zone signs. The Safety Zone sign is used to designate pedestrian refuge and loading islands and used mainly on Melbourne's extensive tram (streetcar) system. The School sign may be used to warn motorists of children on or crossing the road in the vicinity of a school, while the School Zone sign is used in conjunction with speed limit signs which apply when children are going to or leaving school. The Shared Zone sign is essentially a speed limit sign, with the speed limit which applies shown inside a red annulus, with the words, Shared Zone and a depiction of a child and a car for further emphasis. It is applied only at designated shared zones, i.e. zones where the street environment has been modified to encourage low speed travel and a 10 km/h (6 mi/h) limit applies. A rather similar sign is used to indicate Local Traffic Areas, with a pair of child pedestrians similar to those shown on the Children warning sign depicted below the legend, a limit of 40 km/h (25 mi/h) indicated.

8. MIDBLOCK CROSSINGS

Both unsignalized and signalized crossings are provided as mid-block treatments, unsignalized crossings generally being confined to non-arterial roads. The guidelines for installation in AS 1742.10 for an unsignalized crossing specify a shorter critical period of time and lower pedestrian flows, and specify an 85th percentile traffic speed of 80 km/h (50 mi/h) or less.

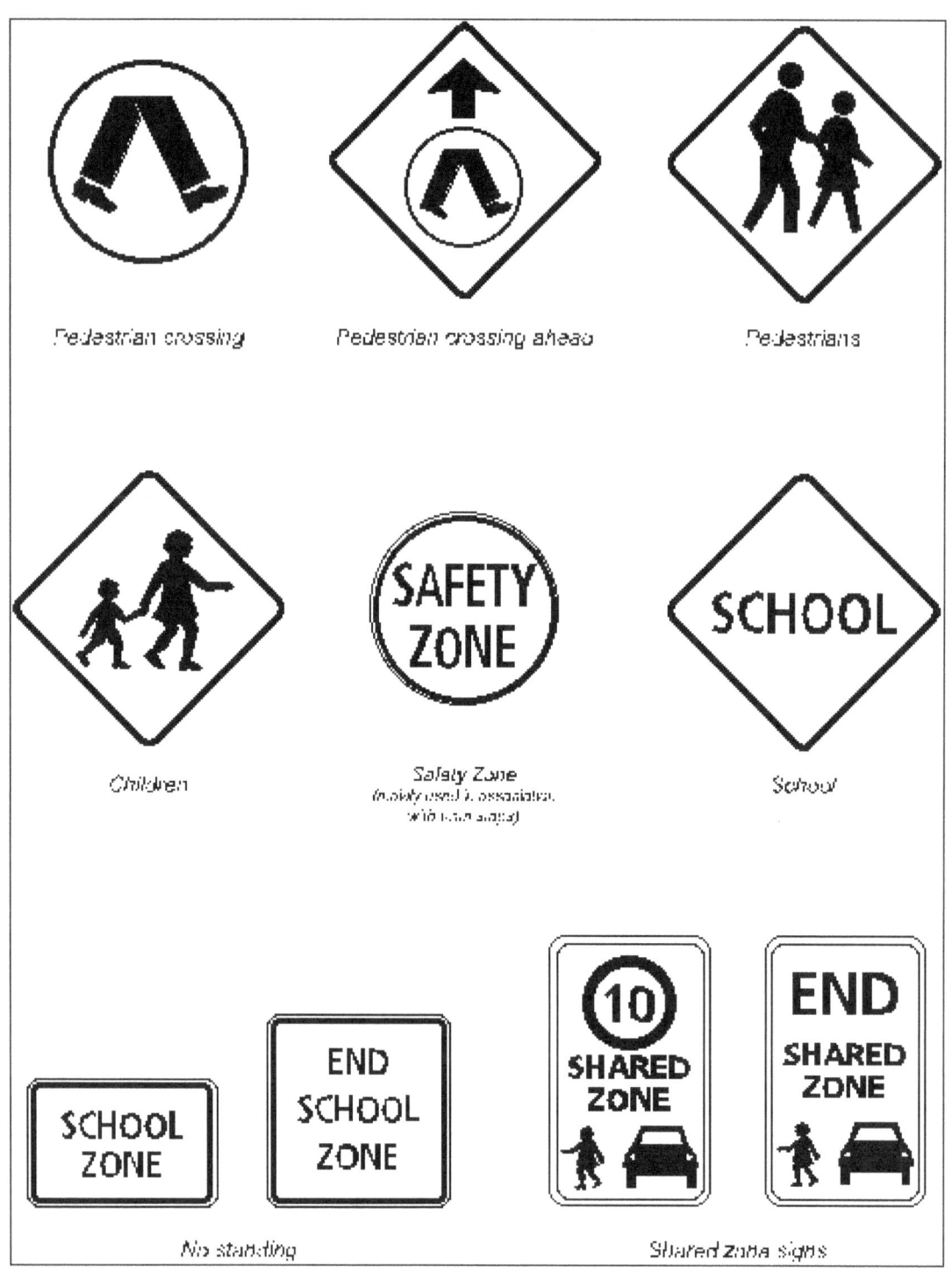

Figure 3. Warning and regulatory signs used in Australia in association with pedestrian facilities or pedestrian activity.

Unsignalized crossings are indicated by painted markings consisting of longitudinal bars approximately 600 mm (23 in) wide with 600 mm (23 in) gaps between, the bars being at least 3.5 m (11.5 ft) long. The markings are generally placed parallel to the center of the road, the crossing may be placed at an angle of up to 30 degrees to suit local circumstances. The crossings are also indicated by the Pedestrian Crossing sign described in section 7, and the Pedestrian Crossing Ahead sign may be used to give advanced warning of the crossing if approaching motorists do not have a clear view of the crossing.

Signalized crossings consist of signal displays facing the drivers in each direction, supplemented by additional high-mounted displays on a mast arm if circumstances warrant, and pedestrian displays as described in section 6 facing the pedestrians. A marked crosswalk and stop lines for the traffic are also provided. Parking is prohibited on the immediate approach to the crossing as required to ensure drivers have an adequate view of the crossing, typically for 3.6 m (12 ft) on either side of the crossing.

Pelican crossings, generally similar to those used in the United Kingdom, have been used in New South Wales and Western Australia for several years, and more recently have been introduced in Victoria. The Pelican crossing is similar to mid-block pedestrian signals, except that during the pedestrian clearance phase, the display facing motorists changes to a flashing yellow, indicating that vehicles may proceed across the crossing, but they are required to give way to pedestrians. Puffin crossings using infra-red sensors to detect the presence of pedestrians and monitor their progress across the crossing have also recently been trialed and are discussed in more detail in section 14.

There appears to be no studies of the safety impact of mid-block Pedestrian Operated Signals (POS) of the type used in Australia. Given the different forms of operation, direct comparison with pelican signals is not appropriate. Fortunately, the Geoplan study included both types of crossing. Installing pelican signals was highly effective in reducing crashes, the quarterly crash rate reducing by 90 percent. Once adjustments were made for reductions at untreated sites, the reduction was 87 percent which was statistically significant. Installing POS also gave rise to statistically significant reductions in crashes. In this case the adjusted reduction was 49 percent.

These results indicate the reductions in crashes brought about by the two types of devices, but they do not reflect the actual safety performance of the devices. To do this would require estimates of the total number of crossings at the different types of devices, making some allowance for different traffic flow conditions. Some indication of the relative safety may be gained by considering the average number of crashes per site per quarter for each type of facility. In this case, the POS had a quarterly rate per site of 0.015 and the pelicans a rate of 0.004. Assuming pedestrian and traffic conditions to be broadly similar at the different types of devices, probably a reasonable assumption in this case, it can be seen that the crash rates are roughly of the same order, with the pelicans apparently having fewer crashes. The average crashes per quarter per site for the intersection signals are 0.05, and for the unsignalized crossings 0.02. It is emphasized again that these are very broad indications and do not adequately reflect the level of risk associated with the different types of facilities.

9. MEDIANS AND PEDESTRIAN REFUGE AREAS

Medians and pedestrian refuges have been widely applied in Australia in recent years. Many roads in urban areas were laid out and built to very generous standards, allowing room for the subsequent installation of a median without compromising traffic flow. Indeed, by providing a turn slot for right-turning traffic, traffic flow was often improved.

Moore and McLean cite early work in New South Wales which showed that the provision of narrow medians reduced vehicle-to-vehicle crashes, but had no effect on pedestrian crashes (Johnston 1962; Leong 1970). However, work carried out in Adelaide, South Australia, suggests that medians are indeed effective in reducing pedestrian crashes. Pedestrian accident rates on arterial roads were found to have an orderly relationship to median width, with the narrowest medians (1.2 m) (4 ft) having four times the pedestrian crash rate of those with the widest median (2.9 m) (10 ft) (Scriven 1986). Replacing a 1.8-m (6-ft) painted median with a wide raised median reduced pedestrian accidents by 23 percent (Claessen and Jones 1994). This reduction is consistent with Scriven's earlier finding that crash rates with 2.9-m (10 ft) raised medians were 33 percent lower than with 1.8-m (6 ft) painted medians.

Support for the safety benefits of refuges islands are less certain. The Geoplan study included four types of pedestrian refuge — those with curb extensions and those without curb extensions, either on existing pedestrian crossings, or on their own. None of them were particularly effective. Refuges with curb extensions actually resulted in an increase in pedestrian crashes; when this was corrected for the reduction in crashes at comparison sites, it resulted in an adjusted crash rate which showed a 53 percent increase. Refuges without curb extensions on existing crossings resulted in no changes in crashes which produced an adjusted rate of a 38 percent increase. Refuges on their own without curb extensions resulted in a 15 percent reduction in crashes, producing an adjusted rate of a 14 percent increase. Only refuges with curb extensions achieved an adjusted rate that was actually a reduction, and that was only 2 percent. However, in view of the other findings discussed above, it seems inherently unlikely that pedestrian refuges did not reduce crashes. The method used in the Geoplan study compared crashes occurring at the site of the facility, before and after. Where pedestrian refuges are provided, it would be expected that pedestrians would be attracted to cross at this point — pedestrians who would otherwise have crossed some distance along the road, so that pedestrian flow is greatly increased at the refuge. A study of the crash history of the whole street where pedestrian refuges have been installed would therefore be necessary to determine whether there had been a reduction in pedestrian crashes.

10. PROVISION FOR THE DISABLED PEDESTRIAN

The Austroads *Guide to Traffic Engineering Practice*, Part 13, draws attention to the problems of providing for people with disabilities and provides guidance throughout the document on the best way to provide for them. Specific topics covered include width of footpaths to accommodate wheelchairs, need for obstruction-free paths, placement of gratings and manhole covers, treatment of ramps and curb ramps, installation of textured paving at waiting areas to provide tactile cues for the visually impaired, loops to detect wheelchairs and allow longer pedestrian green times at signalized crossings, provision of information on routes used by the visually impaired, and signing of facilities and routes for the disabled.

Detailed information on the design of ramps and other features required by the disabled is available in Australian Standard (AS) 1428 — *Design for Access and Mobility*. While new facilities are designed to accommodate the mobility needs of the disabled and much effort has gone into providing ramps and other features on the footpath network, it is not clear how much of the footpath network can be used by different classes of disabled persons, for example, people who walk with sticks may manage stairs but find ramps difficult, in contrast to wheelchair users.

11. SCHOOL ZONE SAFETY

School zone safety is generally addressed by the provision of warning signs to indicate a school zone, and the provision of pedestrian-operated traffic signals or children's crossings, depending on pedestrian and vehicle flows. They may be enhanced by the provision of curb extensions (bulbouts). The children's crossing is a part-time crossing which operates when children are going to or from school. It is indicated by marker posts with red and white alternating bands by the roadside and crosswalk lines painted on the pavement and stop lines 6 m (20 ft) in advance of the crossing. When the crossing is in operation, fluorescent orange flags bearing the legend "Children crossing" are displayed on a post in advance of the crossing, level with the stop line.

Adult crossing supervisors are employed in most States to control the crossings during the times when children are going to or coming from school and traffic flows warrant this. Most signalized crossings close to schools have supervisors on duty, but only some of the children's crossings have supervision. Model instructions for adult supervisors are included in AS 1742.10, including dress and equipment, which includes a white coat, red or orange sash, Stop banner, and a whistle. States generally recognize that the provision of adult crossing supervisors is not a cost-effective road safety measure. For example, in Victoria, crossing supervisors consume about half the budget available for pedestrian safety, despite the fact that only a handful of crashes occur at children's crossings or other crossings where crossing supervisors are present (Klein, personal communication). However, community pressure is such that these programs have continuing support. It would appear that parents regard the crossing supervisor as important for several reasons besides road safety.

South Australia has rather different arrangements. The school warning sign is used to indicate a school zone, in which motorists may travel at no more than 40 km/h (25 mi/h) when children are going to or from school. Wide use is made of a distinctive style of children's crossings provided with flashing lights, travel through the crossing being limited to 25 km/h (16 mi/h) when the lights are operating. Child crossing monitors, operating in teams of three, are used in place of adult crossing supervisors. Model instructions for child monitors are also included in AS 1742.10.

Concern about vehicle speeds on the approaches to school crossings has prompted several States to introduce 40 km/h (25 mi/h) speed zones on a part-time basis. Unlike the situation which applies in South Australia, these are clearly indicated by a 40 km/h (25 mi/h) limit sign. The part-time nature of the sign is catered by either displaying the times at which the limit applies, or having a folding sign that is only opened out when the limit applies. Uber, Barton, and Brown (1992) report on the evaluation of six part-time school speed zones in Victoria, a 40 km/h (25 mi/h) limit being imposed on roads where the prevailing limit was 60 km/h (37 mi/h), and a 60 km/h (37 mi/h) limit on roads with a prevailing limit of 80 (50 mi/h) or 100 km/h (62 mi/h). All zones had a school crossing inside the zone. Folding speed limit signs with the legend "School zone" were installed, with a single flashing yellow light on the pole on the left-hand side of the road. Compared to normal operation of the crossing, operation in conjunction with the part-time speed restriction resulted in considerable speed reductions. With the 60 km/h (37 mi/h) part-time school zone, mean speeds fell by 13 to 19 percent and 85th percentile speeds by 7 to 15 percent in the 80 km/h (50 mi/h) zone, and mean speeds by 30 km/h (19 mi/h) and 85th percentile speeds by 28 to 29 km/h (17 to 18 mi/h) in the 100 km/h (62 mi/h) zones. With the 40 km/h (25 mi/h) limit school zone in 60 km/h (37 mi/h) zones, mean speeds fell by about 20 km/h (12 mi/h) and 85th percentile speeds by 14 to 18 km/h (9 to 11 mi/h). Although the part-time school zones would appear to have been effective in reducing speeds, the authors caution that speeds did tend to creep upwards over a 6-month period and that over half of free speed vehicles were exceeding the school zone limit.

Moore and McLean cite an early literature review (Foldvary 1973) which suggested that school crossings tended to be safer than other types of crossing. This is confirmed by two subsequent Australian studies. Cameron and Jordan (1978) found that the risk to children at a school crossing was approximately one quarter of the risk at a zebra crossing at the same time of day. Bishop and Harwood (1978) compared risk to children at the flashing light school crossing used in South Australia with the risk in school zones (designated in South Australia by a SCHOOL warning sign and an END SCHOOL ZONE warning sign, drivers being limited to 40 km/h (25 mi/h) in the zone when children are entering or leaving school). Risk was significantly lower at the light-controlled crossings.

The Geoplan study, however, indicates that children's crossings may not reduce crashes. Indeed, the installation of childrens' crossings with supervisors actually resulted in an increase in pedestrian crashes. The corrected rate indicated a 77 percent increase in crashes. However, it must be remembered that this data is derived from a small number of sites and an even smaller number of crashes, and that the provision of a crossing with a supervisor would be likely to attract children to the crossing. Installing childrens' crossings without supervisors resulted in a 24 percent reduction in crashes, or a 4 percent reduction following adjustment. This should not be interpreted as unsupervised crossings being safer, as the unsupervised crossings are likely to be at locations where pedestrian and vehicle flows are lower, and the unsupervised crossings are less likely to attract people who would otherwise cross elsewhere.

12. PEDESTRIAN OVERPASSES AND UNDERPASSES

Pedestrian overpasses and underpasses have been provided in Australia in the past but appear to attract little use unless pedestrians have no other means of crossing the road. Currently, their provision would seem to be limited to freeways and other major new projects. There does not appear to be any Australian research on this topic. Daff and Cramphorne (1995, see Section 15) have drawn attention to the need to consider convenience and personal security in the course of conducting pedestrian safety audits. Had safety audit principals been applied in the past, most of the overpasses and underpasses would either not have been built or would have been designed very differently.

13. TRAFFIC CALMING FOR PEDESTRIANS

13.1 Local Area Traffic Management

Local area traffic management (LATM) has been widely adopted in Australia over the last 20 years or so. LATM has been an evolving process, with emphasis shifting from early concern with excluding through traffic from local streets to controlling driver behavior and reducing speeds as the extent of the accident problem on local streets was acknowledged (Brindle 1992). LATM aims to effect these changes by altering the physical environment rather than by regulations and their enforcement. Safety for pedestrians and cyclists, especially safety for young children and old people, is a major concern for people living along local streets and is often an important factor in community support for LATM schemes.

Although part 13 of the Australian standard for traffic control devices is primarily a manual describing the signs and markings to be used with LATM devices and their application, it does contain a table setting out a diagram of each type of physical devices together with its advantages and disadvantages (AS 1742.13: Local Area Traffic Management, Standards Australia 1991). This is intended as a general indication of the range of devices and their appropriate application. Each LATM device has to be designed on an individual basis to take account of site factors, drainage, and so on, and many comprehensive design manuals are available.

There is good reason to expect that LATM would be effective in reducing pedestrian crashes in residential areas. LATM schemes either reduce through-traffic, reducing pedestrian and cyclist exposure to conflict with motor vehicles, or slow vehicles down. As the studies discussed in section 7 show, small changes in speeds greatly improve the outcomes of collisions with motor vehicles for pedestrians. However, tangible improvements to pedestrian or cyclist safety as a result of LATM schemes are elusive. Brindle (1986) reviewed extensive evidence that LATM treatments, both individual devices and area-wide schemes, had been effective in reducing crashes. While the vast majority of studies indicated reductions in crashes, Brindle did not identify any studies that separated crash reductions for pedestrians or cyclists.

Fairlie and Taylor (1990) evaluated the safety benefits of two LATM schemes in Sydney and include breakdowns by crash type (RUM code) and type of unit involved. Although there was a significant reduction in accidents involving cars in one area and a significant reduction in accidents involving motorcycles in the other, the numbers of pedestrian and bicyclist accidents recorded in the areas were too small to come to any conclusions and represent a very small proportion of the accident totals.

It would therefore seem that although a consideration of first principles suggests that LATM ought to be effective in reducing pedestrian and cyclist crashes, the frequency of such crashes is so low, even in local areas, that the benefits are likely to be difficult to detect.

13.2 Effect of humps and raised platforms

While it is intuitively obvious that humps and other measures will reduce traffic speeds, there appear to be few Australian studies on the actual speed reductions associated with humps. Taylor and Rutherford (1986) describe a procedure suitable for measuring the effects of any speed control device, and which they used to assess the effects of road narrowings in local streets. Spot speeds were recorded at points on the approach and departure sides of the slow points. The devices were found to be effective in reducing speeds, vehicles traversing them at speeds for 25 to 30 km/h (16 to 19 mi/h). However, speeds returned to their original level within about 80 m (262 ft) of the slow point. More recently, Zito and Taylor (1996) took advantage of the removal of road humps to measure changes in speed profiles. Speed profiles were obtained using an instrumented car. The profiles show fairly sharp decelerations and accelerations in the vicinity of the humps, from an operating speed of 40 km/h (25 mi/h) to 15 km/h (9 mi/h) or less, and back to 40 km/h (25 mi/h) over a distance of 200 km (124 mi/h) or less.

Such evidence as there is supports the effectiveness of humps and raised pavements as devices that have the capacity to improve pedestrian safety. Moore and McLean cite a study by Jones and Farmer (1993) in which the impact of a series of pedestrian ramps installed along a busy shopping street was assessed. Injury accidents fell from 18 per year to 3 per year, pedestrian delay was reduced, while traffic flow and speed were also reduced. The Geoplan study also included several raised pedestrian platforms. However, it is not clear how many were installed as part of a grouped treatment similar to the treatment studied by Jones and Farmer. Nor is it possible to say how many were installed where there were heavy pedestrian flows. Nevertheless, the results are remarkably similar. A reduction in pedestrian crashes of 94 percent was experienced following installation, in this case giving an adjusted reduction of 93 percent.

Curb extensions, on the other hand, appear to have been relatively successful. Curb extensions on their own produced an adjusted reduction of 27 percent, and curb extensions at existing pedestrian crossings produced an adjusted rate showing a 44 percent reduction.

13.3 Roundabouts (traffic circles)

Australia has made extensive use of roundabouts in recent times, often in association with other LATM measures as well as on arterial roads. They are widely distributed throughout residential areas and so are frequently encountered by pedestrians. Perhaps somewhat optimistically, part 13 of the *Guide to Traffic Engineering Practice* advises that the installation of a roundabout improve pedestrian safety at an unsignalized intersection. The splitter islands on the approaches to the roundabout give pedestrians the opportunity to make staged crossings as does a median or pedestrian refuge. Curb ramps are generally provided 6 m (20 ft) or so away from the entrance holding line, so that pedestrians are encouraged to cross to the rear of the first car waiting to enter the roundabout, thus reducing the risk of not being seen by a driver concentrating on finding a gap in the traffic on the roundabout. If a pedestrian crossing (zebra) is required across one of the entry legs of the roundabout, it should be located at least 12 m (40 ft) from the exit.

Although roundabouts are recognized as a treatment that is effective in reducing the severity of crashes, there does not appear to be Australian data on their effect on pedestrian crashes.

14. INNOVATIVE DEVICES

14.1 Infra-red sensors

As described in section 8, Victoria has proceeded with trial installations of a version of the UK Puffin crossing. It is similar to the UK crossing in that it uses infra-red detectors to monitor the progress of pedestrians across the road, but differs in that infra-red detectors rather than pressure mats are used as pedestrian presence detectors. It also differs in that conventional call buttons and pedestrian signal displays are used, rather than call button and low-mounted display visible only in the pedestrian waiting area. The system is designed for easy conversion of existing mid-block signals.

Catchpole, Jordan, and Cairney (1996) evaluated the behavior of drivers and pedestrians at a trial installation of a Puffin crossing in Victoria. This involved conversion of a standard Pedestrian Operated Signal (POS) site to Puffin operation. Pedestrian green was changed from a fixed period of 8 seconds to a minimum of 4 seconds and a maximum of up to 10 seconds. A 40-percent reduction in vehicle delays was evident, and there was no increase in red running or other driver behaviors that might adversely affect safety. There was an increase in pedestrian compliance with the signals. There was a significant reduction in the percentage of pedestrians starting to cross before the green (10%), and non-significant increase in persons crossing on the green signal. As would be expected with a shorter fixed green time, more pedestrians completed their crossing during the flashing red period under Puffin operation, but there were no more on the crossing at the start of the steady red period than had been the case under the steady red period. Thus Puffins appear to deliver a longer crossing time to pedestrians who need it and reduce delays to motorists without compromising pedestrian safety.

14.2 Pelican crossings

Pelican crossings have been extensively used in the UK and have been in use in some parts of Australia for many years. It is probably a fair generalization that they are seen as a compromise solution that allows pedestrians crossing opportunities, but which minimize delay to traffic. Concern has been expressed about the large proportion of pedestrian crashes which occur at Pelican crossings (e.g., Hunt 1992). The study of installations in Sydney (Geoplan 1994 - see section 8) showed that Pelicans were an effective way of reducing pedestrian crashes. However, the very fact that they allow traffic to proceed through the crossing during the clearance phase can be used to pedestrians' advantage as under these circumstances motorists will readily tolerate double cycling, i.e., the introduction of the pedestrian phase twice in the normal cycle. This results in more opportunities for pedestrians to cross while traffic is prevented from entering the crossing, yet allows traffic to move more freely than it does under the normal mid-block signal arrangements. The low cost of conversion of mid-block signals to Pelican type operation, and the even lower cost of introducing the double cycling option is likely to prove very appealing to road authorities.

Observing pedestrian behavior at two crossings before and after conversion to double-cycle Pelican operation, Daff and Cramphorne (1995b) found that fewer pedestrians crossed away from the crossing, and that fewer crossed against the steady red "Don't Walk" indication. Drivers welcomed the reduction in what they perceived as unnecessary delay with normal mid-block signal operation.

14.3 Illuminated push buttons

Illuminated push buttons have been used in Australia for many years. The earlier type of push button has a small horizontal panel that lights up to display the words "Call recorded." These were replaced with a more robust type of switch that contained no illumination. However, these push buttons have been superseded by a similar type that contains a small red circular display illuminated by LEDs when the switch is pressed. The process has been one of evolution and has not proved to be controversial.

It is also worth noting that push button units with an auditory signal for the benefit of sight-impaired pedestrians are widely used. During the pedestrian clearance interval and "Don't Walk" interval, a slow beat is emitted. On the signals changing to the "Walk" interval, a brief, distinctive high-pitch signal is emitted, followed by a fast steady beat, reverting to the slow beat at the start of the pedestrian clearance interval. In addition to the auditory signal, a tactile pulse is provided through the central button on more recent models.

Requirements for the design, construction, and performance of push-button assemblies for pedestrian crossings are set out in Australian Standard (AS) 2353 (Standards Australia, 1997). At time of writing, a new draft standard had been issued for public review. The aim of the revision is to introduce more comprehensive requirements for the auditory and tactile signals.

15. OTHER ISSUES

15.1 Multi-action programs

The concept of multi-action programs evolved in VicRoads (Victoria's State Road Authority) as a response to the usual situation of there being no one intervention or narrow range of actions that is likely to have a major impact on pedestrian safety. The immediate background to this concept is a series of investigations carried out by Monash University Accident Research Centre on behalf of VicRoads to identify crash patterns and contributing factors at locations with concentrations of pedestrian crashes (Corben and Diamantopoulou 1996). This research suggested that a single countermeasure or limited range of countermeasures was likely to be of limited effectiveness in reducing crashes, it was in response to these findings that VicRoads staff conceived an integrated package embracing road and traffic engineering, education, publicity, and enforcement which constitutes the WalkSafe program, as the multi-action program has come to be known.

Conceptually, the approach begins with an accident investigation and safety audit to understand the nature of the problem and contributing factors. Such audit also aids the design of countermeasures and ensure that physical treatments and enforcement are appropriately located. This is followed by implementation — installation of engineering treatments, delivery of education and publicity, and enforcement of traffic regulations focused at key locations. Implementation is accompanied by evaluation. Given the relatively low numbers of pedestrian crashes, this generally has to be in terms of evaluating behavioral changes rather than actual crash reductions, at least in the short term.

At present, the first multi-action program is beginning in Stonnington, a municipality close to Melbourne's central business district. A budget of more than $A 1 million dollars (approximately $US 600,000) has been allocated to the program, and extensive evaluation of the program is currently being undertaken (Klein, personal communication). Physical treatments being trialed include a painted median placed between the tram tracks.

15.2 Community Road Safety

Several states have been active in developing community road safety over the last few years. Although the organization supporting community road safety and its funding mechanisms vary from State to State, community road safety programs have the following features in common:
- They deal with issues at a local level.
- They encourage community ownership of road safety problems and seek community involvement in their solution.
- Solutions are customized to fit the needs of individual communities.

New South Wales has the most complete program in Australia, with Road Safety Officers employed by several Local Councils and with good support in terms of advice on effective programs, materials, and training (St. John, Clarke, and Graham 1996). Victoria also has paid Road Safety Officers, who support Community Road Safety Committees, which typically cover several local government areas. As discussed above, it also has pedestrian advocates who are very involved with community projects. In Western Australia, a small unit operating as part of the Western Australian Municipal Association has responsibility for promoting road safety activities among local communities, providing advice, materials, and a limited amount of funding. Some community road safety activity has been evident in other States, but as yet they have no formal delivery mechanism.

The principles advocated for community road safety programs are almost identical to those advocated for the multi-action programs: analysis of the problems based on objective data and community perceptions, integrated and targeted actions to address the most pressing problems, and evaluation and review to ensure intended actions are carried out effectively and that the desired changes are being achieved. Although community road safety programs address a range of problems besides pedestrian safety, experience to date suggests that pedestrian issues are very high on community agendas.

Many municipalities, especially in NSW, have developed community road safety plans. Typically, issues of pedestrian and cyclist safety feature largely in these, especially those involving young or elderly road users. The solutions which the plans attempt to implement cover a range of actions, from provision of bicycle and pedestrian crossing facilities to educational campaigns and promotions. At time of writing, no formal evaluations of the programs had been published, although reviews in NSW and Western Australia were in progress.

Community road safety appears to be a promising way of delivering improved safety to pedestrians and cyclists. However, there is no evidence as yet that this approach is effective in reducing crashes.

15.3 Pedestrian safety audits

The principles of road safety auditing are by now well established in Australia (Austroads 1994), although much of the road network remains to be audited and it is at times unclear what the priorities for remedial action should be. An audit, which should be carried out by a qualified, independent assessor, is a formal assessment of the accident potential and likely safety performance of an existing or future project. One of the key principles of the audit process is for the auditor to view the facility from the perspective of the road user, and this applies to pedestrians and cyclists as much as to any other type of road users.

Jordan (1995) describes how the Austroads audit process might be used to address the particular needs of pedestrians. Safety audit ideally should happen at several stages in the development of a facility, and there are characteristic issues to be dealt with at each stage:

- Feasibility stage: who are the likely users, what standards should be used.

- Layout design stage: pedestrian routes and facilities in relation to horizontal and vertical alignments and other features of the proposal.

- Detailed design stage: can the pedestrian see and be seen adequately, signing, lighting, signal timings, likely use by large numbers of users with special needs.

- Pre-opening stage: has the job been completed or designed, and do any changes affect pedestrian safety? Can pedestrians cope well with the facility provided? Night-time inspection is particularly important.

- Network review (audit of existing facilities): are safety features consistent with its functional classification, are they adequate for any changed circumstances, and have there been changes to the site which pose safety problems.

Jordan briefly cites a number of examples where safety audits have revealed problems for pedestrians, and which were able to be remedied before the facilities in question were opened to the public. He concludes with a consideration of the needs of all pedestrians as well as pedestrians with special needs — the elderly, the young and the intoxicated.

A rather broader view of pedestrian audits is put forward in the same conference proceedings by Daff and Cramphorne (1995), who envisage a pedestrian audit as embracing issues relating to convenience, mobility, and personal safety. They give examples of points that might be considered in each of these check lists and examples of the sorts of recommendations likely to be made as the result of a pedestrian audit. Among their suggestions for future development of the audit process are a consideration of behavioral observations as to how these facilities are used, and the use of archival material such as records of accidents or police records of assaults.

Although the outcome of the process is a set of recommendations rather than an index describing the quality of the pedestrian experience, the process and the nature of the judgments and the specific features of the environment considered, have much in common with the level-of-service measures for pedestrian facilities proposed by Khisty (1994) and Sarkar (1995), which may be familiar to North American readers.

15.4 Speed and pedestrian safety

Recent Australian work reinforces current understanding of the sensitivity of the outcome of pedestrian crashes to small changes in vehicle operating speeds in the range typical of local urban streets. McLean et al (1994) applied accident reconstruction techniques to estimate the probable speeds of vehicles involved in pedestrian fatal crashes in the Adelaide Metropolitan area. Most of the crashes occurred on speeds with 60 km/h (37 mi/h) speed limits. They estimated the likely outcomes of the crash under several different scenarios, including one where no vehicle exceeded the speed limit and one where all vehicles were traveling a uniform 5 km/h (3 mi/h) slower than they actually had been at the time of the crash. Under this last scenario, 32 percent of the pedestrians would have survived the impact, and a collision would have been avoided altogether in 10 percent of cases. Eliminating travel above the existing speed limit would have been considerably less effective than a uniform reduction of 5 km/h (3 mi/h).

These findings have provided the inspiration for some very dramatic road safety advertising developed by Victoria's Transport Accidents Commission, a body which integrates third party injury insurance with rehabilitation and crash prevention. This advert graphically depicts a pedestrian being thrown in the air when the vehicle approach speed is 70 km/h (43 mi/h), compared to a narrowly averted impact when the approach speed is 60 km/h (37 mi/h).

15.5 Impact of lower speed limits on pedestrian safety

While there is widespread agreement that lower vehicle speeds will reduce crashes, including crashes involving pedestrians and cyclists, it is difficult to specify the precise nature of the relationship between speed reductions and crash reduction. The extent to which lower speed limits will be effective in reducing speeds add another element of uncertainty in the process of making predictions. The evidence showing a reduction in crashes in response to lower speed limits and lower speeds has been covered in a recent review of urban speed management in
Australia (Austroads 1996). That report pointed out that, with a general urban speed limit of 60 km/h (37 mi/h), Australian urban speed limits are high by comparison with other developed countries.

Australia has had little experience with lower urban limits, and early experience suggested that lower limits were only effective in conjunction with physical devices designed to slow traffic. Although a more recent trial has suggested that speeds on local streets can be reduced by speed limits alone when accompanied by modest levels of enforcement, it is not known how effective a system-wide change to speed limits on local streets would be. The report suggested that a 50 km/h (31 mi/h) limit would be suitable for local streets, representing a reduction in speeds that most motorists would be prepared to accept. The report emphasizes the need for States to consider any changes to local street speed limits as a package which embraces publicity and

education, community consultation, and enforcement, with monitoring and evaluation of changes to vehicle speeds, accidents, travel times, and amenity.

15.6 The impact of seat belt wearing legislation on pedestrian safety in Australia

Although there is no doubt that restraint wearing does reduce deaths and injuries to vehicle occupants, there has been some controversy over the indirect impact they may have on pedestrian safety. As Australia was very much a pioneer of compulsory restraint-wearing laws, its experience in regard to this controversy is of particular interest.

It has been argued that one effect of restraint wearing by drivers is to increase their feelings of safety, which they then trade-off through reduced gap-acceptance, increased speeds and so on, so that they maintain a constant level of risk (eg. Wilde 1986). Adams (1985) has argued that this is likely to have adverse safety consequences for other road users. Any increase in vehicle speeds is likely to increase the risk and severity of collisions with pedestrians. As we saw in chapter 7, small changes in vehicle speeds can have very great consequences for the outcomes of collisions with pedestrians.

Connybeare (1980) demonstrated that in Australia, in the years following the introduction of compulsory restraint-wearing laws, vehicle occupant deaths were well below the levels that would be predicted from a linear regression model based on the pre-seat belt years, and that non-occupant deaths were above the figures predicted by the model. He then made the unfortunate conclusion that this put auto safety measures in the class of self-defeating class of government policies which show no net benefit in welfare. As Hampson (1982) clearly pointed out, this conclusion is flawed in that Connybeare neglected to differentiate between different classes of non-occupant, which included motor cycle riders and passengers, as well as pedestrians and cyclists, and that the period when seat belt legislation was introduced throughout Australia was a period of increasing motor cycle ownership and use. Examining the disaggregated trends in New South Wales over the years 1961 to 1980, Hampson shows a rising trend in motorcyclist fatalities from 1960 to 1971, a sharp increase in the period 1970 to 1975, then a leveling off. Fatalities involving pedestrians and other non-occupants declined steadily over the period. Thus there would appear to be no grounds for supposing that there was an increase in pedestrian deaths in NSW following the introduction of compulsory restraint wearing laws. As was seen in section 2, despite very high restraint wearing rates, pedestrian fatalities and injuries have continued to decline in Australia in recent years. However, this may be attributable in part to the decline of walking as a transport mode.

16. EDUCATIONAL CONSIDERATIONS

All Australian States and Territories have elements of road safety available for teaching in primary and secondary schools. However, the extent to which road safety is taught is left to individual schools to decide, with some having no road safety, and others making a considerable investment in it, including hands-on driver training in some cases.

Victoria is among the leaders in developing educational materials for use in schools and has published thorough-going evaluations of its programs from pre-school through to secondary school. Anthony and Wilcock (1991) found that 55 percent of early childhood centers used Starting Out Safely, a program aimed at developing safe pedestrian and restraint use behaviors for pre-schoolers. The main reasons for not using the programs were pressure of other activities, the cost of the program, or the centers being unfamiliar with the program.

Seventy-eight percent of primary schools included traffic safety education in the curriculum (Anthony, Cavallo, and Crowle 1992). The percentage was highest in large rural schools, and lowest in nongovernment schools in the metropolitan area. The main reasons for NOT providing traffic safety education were other demands of the curriculum, or lack of trained and interested staff. Forty six percent were using "Streets Ahead" or its predecessor, programs aimed at safe pedestrian behaviors, and 38 percent used "Bike-Ed", aimed at developing safe cycling behavior. Some difficulties were reported with Bike-Ed, mainly the problem of finding enough trained adult supervisors for the onroad cycling components.

Traffic safety education was also widely taught in secondary schools, with 88 percent of schools using at least one of the traffic safety lessons developed by VicRoads, and 80 percent using two or more (Anthony et al 1992). The most popular units were one on alcohol and driving, and a legal studies unit relating to car ownership and operation. Thirty-nine percent of schools conducted safe cycling programs. At this level the focus would therefore appear to be on preparation for driving rather than safe pedestrian behaviors.

A more recent study of traffic education in general in Victoria indicated that primary schools spent an average 86.8 hours teaching road safety, compared to an average of 57.9 hours in secondary schools (Harrison, Penman, and Penella 1997). The use of resources varied considerably between years and grade levels. A minority of teachers had negative attitudes toward traffic safety education. Of those who did have negative attitudes, this was largely because of time pressure or an unfavorable view of the materials.

Safe routes to school programs are now firmly established in at least three States. The essence of the programs is a package of measures aimed at reducing the risk encountered in the course of journeys to school through a package of integrated activities, including the promotion of the safest routes, the provision of some low-cost engineering treatments to reduce risk at hazardous locations, and education of the broad school community in the philosophy behind the route and its safe use. Details of the process, funding arrangements, and programs differ from State to State.

In Australia, there appears to be little formal documentation of what has been achieved under safe routes to schools programs. Healy (1995) gives a brief account of the program as it has developed in Victoria. A key driver of the program is the Local Pedestrian Advocate, a member of the road authority's regional staff whose duties include identifying locations with concentrations of accidents involving young children and marketing the program to schools and municipalities. Once a school decides to take on a safe routes to school program, the steps in implementation are:

- A home survey to identify routes and points of difficulty children experience as pedestrians and cyclists.
- Matching survey results with accident data.
- On-site observation.
- Presentation of recommended action to the school and local community before implementation of measures.
- Support for teachers in planning activities to encourage children to adopt safer travel behaviors.

Healy reports that at time of writing, the program had been implemented in 12 municipalities with the participation of over 100 primary schools. However, an evaluation of the program had concluded that crash numbers during the before and after periods were too small to allow statistically reliable conclusions (Tziotis 1994, cited in Healy 1995).

Other States have adopted similar programs, but there appears to be no definitive evidence that the programs have been effective in reducing crashes. However, safe routes to school have an intuitive appeal and are very effective in achieving a community focus in road safety.

16.1 Walk with care

Walk with care is a program provided throughout Victoria which is aimed at improving the safety of elderly pedestrians. Healy (1995) gives a succinct overview of the program. The approach adopted is similar in some respects to that adopted in the Safe Routes to School program in that it focuses on high accident locations, encourages community ownership of the problem, and relies on an integrated approach to tackling well defined local problems. The essential elements of the program are:

- Specially trained volunteer discussion leaders conduct sessions with groups of older residents throughout a municipality to give information about safe ways of using roads and collect information on problem spots.

- A survey is distributed to older persons throughout the municipality to supplement the discussions.

- Council staff determines necessary engineering works on the basis of discussion group outcomes, survey results, and crash statistics.

- Engineering works and safety messages are publicized through local newspapers, information bulletins, and municipal newsletters.

As was the case with Safe Routes to School, an evaluation determined there was insufficient crash data to arrive at reliable conclusions (Tziotis 1994, cited in Healy 1995). In view of the difficulty in evaluating such programs by their effects on crash reductions, several alternative means of evaluating future programs have been suggested. These include giving leaflets with information about correct crossing procedures to people crossing the road incorrectly, followed 1 week later by enforcement of pedestrian regulations at the same site, video recording of crossing behaviors, and continuing group discussions at three monthly intervals (Robins, personal communication).

17. ENFORCEMENT AND REGULATION

The use of enforcement to achieve greater compliance with traffic regulations on the part of pedestrians and cyclists has generally not attracted sustained effort on a large scale by police forces in Australia or elsewhere. Very little literature appears to have been published on the subject, and it is a telling fact that in the latest comprehensive review of traffic law enforcement, running up to 163 pages, pedestrian and cyclist issues do not warrant even a mention or a short section describing the problems associated with this type of enforcement (Zaal 1994).

It is fairly evident why enforcement of pedestrian and bicycle regulations are not particularly attractive to police and probably do not constitute a good use of police resources. These would include:

- Low monetary fines for most offenses, hence perceived low importance.

- Many offenses, especially bicycle offenses, are committed by children, can be time consuming to deal with, and may not involve any monetary penalty.

- Pedestrians and cyclists put themselves rather than others at risk when they disobey regulations.

- Many of the offenses require a lot of police time to detect, either because they occur with very low frequency at any particular place or the presence of police deters their occurrence.

- They are not offenses that are amenable to the type of automated enforcement techniques which have revolutionized the policing of speeding, alcohol, and some other types of violations.

18. SUMMARY AND DISCUSSIONS

Australia now has a relatively good traffic safety performance and, on a population basis, has a lower pedestrian fatality rate than most other developed countries. However, pedestrian crashes continue to be a major and unacceptably high source of deaths and injury. Although this low pedestrian fatality rate is caused by some extent to the heavily car dependent life-style of most Australians, Australia would appear to cater reasonably well for pedestrians in terms of the provision of facilities, driver awareness of and care for pedestrians, and provision of pedestrian programs. It is another question, however, as to whether it could be said to cater well for pedestrians in terms of planning decisions which encourage car use, generate expansion of the urban areas and thereby perpetuate low urban densities that tend to eliminate walking as a viable mode for many trips.

In terms of the pedestrian devices available, Australia would seem to have a full range of treatments available. While major pedestrianization of central city areas has generally been fairly limited, Australia has been a pioneer of traffic calming in the form of Local Area Traffic Management, particularly in residential areas. However, it is not clear that this has produced large benefits in terms of pedestrian safety. Some innovation has been evident in the traffic signals area. Puffin crossings equipped with infra-red detectors to allow slow pedestrians time to complete their crossings look promising although the high cost is likely to limit their application to areas of high need. Pelican crossing are likely to find ready application, and having them set-up for double-cycle operation appears to offer benefits in terms of service to pedestrians, and greater use of the crossings and conformity with signals.

The process of safety audit is now well established in Australia and likely to ensure that new facilities are safer for pedestrians and to reduce the need for costly changes to rectify safety problems once facilities have been built. The process of auditing existing facilities is more problematical, and remedial action for all but the most pressing problems is likely to take some considerable time.

Good teaching materials have been developed to support pedestrian education and other road safety matters in schools, but use of these materials is unevenly spread across the school system. The amount of time and resources dedicated to road safety education is essentially a matter for individual schools. Where Australia has been particularly innovative is in developing the "Safe routes to school" programs, an approach that integrates education with mode and route selection and engineering treatments to develop safe travel behavior and in which pedestrian issues feature prominently. Also worthy of note, although a less fully developed program, is the "Walk with care" program that is designed to raise awareness of pedestrian safety issues among the elderly.

Multi-action programs would appear to be a promising method for addressing pedestrian problems at a local level, and provide a vehicle for integrating many of the strategies and countermeasures discussed above.

19. REFERENCES

1. ADAMS, J. (1985). "Smeed's Law, Seat Belts, and the Emperor's New Clothes." In *Human Behaviour and Traffic Safety*, pp 193-257. New York : Plenum.

2. ALEXANDER, K., CAVE, T., and LYTTLE, J. (1990). *The Role of Alcohol and Age in Predisposing Pedestrian Accidents.* Roads Corporation (Victoria - now VicRoads) Report No. GR/90-11.

3. ANDERSON, P.R., MONTESIN, H.J., and ADENA, M.A. (1989). *Road Fatality Rates in Australia 1984-85.* Federal Office of Road Safety, Report CR 70.

4. ANDREASSEN, D.C. (1992). *Costs for Accident-Types and Casualty Classes.* Australian Road Research Board Ltd, Research Report ARR 227.

5. ANDREASSEN, D.C. (1994). *Model Guidelines for Road Accident Data and Accident-Types Version 2.1.* Australian Road Research Board Technical Manual ATM 29.

6. ANTHONY, S. AND WILCOCK, C. (1991). *Traffic Safety Education in Victoria. Volume 1 - Early Childhood Centres 1990.* VicRoads Report GR 91-12.

7. ANTHONY, S., CAVALLO, A., AND CROWLE, J. (1992). *Traffic Safety Education in Victoria. Volume 12 — Primary Schools 1990.* VicRoads Report GR 92-1.

8. ANTHONY, S., ALLEN, P., CAVALLO, A., AND HARALAM, H. (1992). *Traffic Safety Education in Victoria. Volume 3 — Secondary Schools 1990.* VicRoads Report G 92-2.

9. AUSTRALIAN BUREAU OF STATISTICS (1995). *Australian Year Book 1995.* Canberra: Australian Bureau of Statistics.

10. AUSTROADS (1994). *Road Safety Audit.* Sydney : Austroads.

11. AUSTROADS (1995). Proposed Australian Road Rules ; draft for public comment. National Road Transport Commission.

12. AUSTROADS (1996). *Urban Speed Management in Australia.* Austroads National Report AP118.

13. BISHOP, R.M. and HARWOOD, C.J. (1978). *School Children Pedestrian Measures and Accidents in South Australia.* Proceedings of the Joint ARRB/ DoT Pedestrian Conference, Sydney, November 15-17th 1978, Session 3, Paper 1. Vermont South: Australian Road Research Board.

14. BRINDLE, R.E. (1986). "The Relationship Between Traffic Management, Speed and Safety in Neighbourhoods." Proceedings of the 13th Australian Road Research Board Conference 13, (9), pp 90-102.

15. BRINDLE, R.E. (1992). Australia's Contribution to Traffic Calming, PTRC Summer Annual Meeting, Manchester, September 1992. Proceedings of Seminar G, Traffic Management and Safety, pp 49-60.

16. CAIRNEY, P.T. (1988). *Should the Flashing Man be Red or Green?* Australian Road Research, 18(1), pp 38-40.

17. CAIRNEY, P.T. *Understanding Traffic Control Devices*, Australian Road Research Board, Special Report No. 44, 1989.

18. CAIRNEY, P. AND CUSACK, S. (1997). *Comparison of Pedestrian and Bicycle Accidents in New South Wales, Victoria and Queensland.* ARRB Transport Research Contract Report CR OCN517.

19. CAMERON, M. and JORDAN, P.W. (1978). *Pedestrian Accident Risk at School Crossings and the Effect of Crossing Supervisors in Victoria.* Proceedings of the Joint ARRB/ DoT Pedestrian Conference, Sydney, November 15-17th 1978, Session 6, Paper 3. Vermont South: Australian Road Research Board.

20. CATCHPOLE, J.E., JORDAN, P.W. and CAIRNEY, P.T. (1996). "Mid-block Pedestrian Signals: Puffins and Pelicans Taking Off in Victoria." *Proceedings of the 18th ARRB Transport Research Conference*, 18 (5), pp 405-422.

21. CATCHPOLE, J.E., MILLAR, L., and MISSIKOS, A. (1996). "A More Intuitive Signal for the Pedestrian Clearance Interval. Mid-block Pedestrian Signals: Puffins and Pelicans Taking Off in Victoria." *Proceedings of the 18th ARRB Transport Research Conference*, 18 (5), pp 373-390.

22. CLAESSEN, J.G. and JONES, D.R., (1994). "The Road Safety Effectiveness of Raised Wide Medians." *Proceedings of the 17th Australian Road Research Board Conference* 17, (5), pp 269-287.

23. CONNYBEARE, J.A.C. (1980). "Evaluation of the Automobile Safety Regulations: The Case of Compulsory Seat Belt Legislation in Australia." *Policy Sciences*, 12, pp. 27-39.

24. CORBEN, B. and DIAMANTOPOULOU, K. (1996). *Pedestrian Safety Issues for Victoria.* Monash University Accident Research Centre, Report No. 80.

25. DAFF, M. AND CRAMPHORN, B. (1995A). "Pedestrian Audit: A Process to Raise the Consciousness of Designers." *Proceedings from the 1994 Australian Pedestrian and Bicyclist*

Safety and Travel Workshop, pp 223-230.

26. DAFF, M. AND CRAMPHORN, B. (1995B). "Monitoring the Conversion of Pedestrian-Operated Signals to Double Cycling Pelican Operator." *Proceedings from the 1994 Australian Pedestrian and Bicyclist Safety and Travel Workshop*, pp 301-311.

27. DEPARTMENT OF TRANSPORT (1996). *Road Accidents Great Britain 1995 : The Casualty Report*. London : Her Majesty's Stationery Office.

28. FAIRLIE, R.B. and TAYLOR, M.A.P. (1990). "Evaluating the Safety Benefits of Local Area Traffic Management." *Proceedings of the 15th Australian Road Research Board Conference*, 15 (7), pp 141-166.

29. FOLDVARY, L.A. (1973). *A Review of Pedestrians, Pedalcyclists and Motorcyclists in Relation to Road Safety*. Canberra: Australian Government Publishing Service, Report No. NR/19.

30. FRAINE, G. (1995). *Pedestrian Safety and Travel in Queensland: Issues paper*. Land Transport and Safety Division, Queensland Transport.

31. GEOPLAN (1994). *Evaluation of Pedestrian Road Safety Facilities*. Report prepared for the Road Safety Bureau, Roads and Traffic Authority (New South Wales) by Geoplan Urban and Traffic Planning.

32. HAMPSON, G. (1982). "The Theory of Accident Compensation and the Introduction of Compulsory Seat Belt Legislation in New South Wales." *Proceedings of the 13th ARRB Conference*, 11(5), pp 135-140.

33. HARRISON, W.A., PENMAN, I., AND PENNEELLA, J. (1997). Monash University Accident Research Centre Report 110.

34. HEALY, D. (1995). *Community Involvement in Road Safety-Victoria*. Proceedings of the Australian Institute of Planning and Management Victoria State Institute, April 1995.

35. HUNT, J. (1992). "The Operation and Safety of Pedestrian Crossings in the United Kingdom." *Proceedings of the 17th Australian Road Research Board Conference*, 17 (5), pp 49-64.

36. JORDAN, P. (1995). "Road Safety Audit — What It Can Do to Improve Safety for Pedestrians." *Proceedings from the 1994 Australian Pedestrian and Bicyclist Safety and Travel Workshop*, pp 213-222.

37. JOHNSTON, R.E. (1962). "Experience with Narrow Medians." *Proceedings of the 1st Australian Road Research Board Conference*, 1 (1), pp 489-499.

38. JONES, S.M., and FARMER, S.A. (1993). "Pedestrian Ramps in Central Milton Keynes: A Case Study." *Traffic Engineering and Control*, 34 (3), pp 122-128.

39. KHISTY, C.J. (1994). "Evaluation of Pedestrian Facilities: Beyond the Level-of-Service Concept." *Transportation Research Board. No. 1438*, pp 45-50.

40. KLEIN, R. Personal communication.

41. LEONG, H.J.W. (1970). "The Effect of Kerbed Median Strips on Accidents on Urban Roads." *Proceedings of the 5th Australian Road Research Board Conference*, 5 (3), pp 338-364.

42. MCLEAN, A.J., ANDERSON, R.W.G., FARMER, M.J.B., LEE, B.H., and BROOKS, C.G. (1994). *Vehicle Travel Sspeeds and the Incidence of Fatal Pedestrian Collisions, Volume 1.* Federal Office of Road safety, Report No. CR 146.

43. MOORE, V.M., and McLEAN, A.J. (1995). *A Review of Pedestrian Facilities*, Office of Road Safety, South Australian Department of Transport, Report Series 2/95.

44. NHTSA (1996). *Traffic Safety Facts.* Washington : National Highway Traffic Safety Administration.

45. ROBINS, M. Personal communication.

46. ROSMAN, D.L., and KNUIMAN, M.W. (1994). "A Comparison of Hospital and Police Road Injury Data." *Accident Analysis and Prevention*, 26 (2), pp 215-22.

47. RTA (1994). *Road Traffic Accidents in New South Wales 1993.* Sydney; Roads and Traffic Authority.

48. SARKAR, S. (1995). "Evaluation of Safety for Pedestrians at Micro- and Micro-Levels in Urban Areas." *Transportation Research Record No. 1502*, pp 105-108, 1995.

49. SCRIVEN, R.W. (1986). "Raised Median Strips — A Highly Effective Road Safety Measure." *Proceedings of the 13th Australian Road Research Board Conference*, 13 (9), pp 46-53.

50. STANDARDS AUSTRALIA (1993). AS 1428, *Design for Access and Mobility.* North Sydney: Standards Australia.

51. STANDARDS AUSTRALIA (1990). AS 1742.10, *Manual of Uniform Traffic Control Devices, Part 10: Pedestrian Control and Protection* North Sydney: Standards Australia.

52. STANDARDS AUSTRALIA (1991). AS 1742.13, *Manual of Uniform Traffic Control Devices, Part 13: Local Area Traffic Management*. North Sydney: Standards Australia..

53. STANDARDS AUSTRALIA (1997). Draft Australian Standard — Pedestrian Push-Button Assemblies. Homebush: Standards Australia DR 97515 (Revision of AS 2353).

54. St. JOHN, L., CLARKE, J., and GRAHAM, A. (1996). "Local Government Road Safety Program: Building Capacity, Achieving Effectiveness and Having an Impact." *Proceedings of the 18th ARRB Transport Research Conference*, 18 (5), pp 361-371.

55. TAYLOR, M.A.P., and RUTHERFORD, L.M. (1986). "Speed Profiles at Slow Points on Residential Streets." *Proceedings of the 13th Australian Road Research Board Conference*, 13 (9), pp 65-77.

56. TZIOTIS, M. (1994). "Pedestrian Crashes in Melbourne's Central Activity District." *Proceedings from the 1994 Australian Pedestrian and Bicyclist Safety and Travel Workshop*, pp 99-112.

57. UBER, C.B., BARTON, E.V., and BROWN, F. McK. (1992). "Trial of Part-Time School Speed Zones." *Proceedings of the 16th ARRB Conference, Perth*, Vol 16 (4), pp 203-18.

58. WILDE, G.J.S. (1986). "Beyond the Concept of Risk Homeostasis: Suggestions for Research and Appreciation Towards the Presentation of Accidents and Lifestyle-Related Disease." *Accident Analysis and Prevention*, 18(5), pp 377-401.

59. ZAAL, D. (1994). *Traffic Law Enforcement: A Review of the Literature*, Federal Office of Road Safety, Report No. 53.

60. ZITO, R., and TAYLOR, M.A.P. (1996). "Speed Profiles and Fuel Consumption at LATM Devices." *Proceedings of the 18th ARRB Transport Research Conference*, 18 (5), pp 391-406.

www.ingramcontent.com/pod-product-compliance
Lightning Source LLC
Chambersburg PA
CBHW081400170526
45166CB00010B/3153